EMERGENCY!

EMER

For Ann and Michael
MM

For Marcus, Michelle and Jay
AA

ISBN 0-439-67644-4

12 11 10 9 8 7 7 8 9/0

Printed in the U.S.A. 40

First Scholastic printing, September 2004

GENCY!

written by Margaret Mayo

illustrated by Alex Ayliffe

SCHOLASTIC INC.

New York Toronto London Auckland Sydney
Mexico City New Delhi Hong Kong Buenos Aires

Call 911 - emergency!
Burglars make a getaway.
Police car dashing, bright lights flashing,
Help is coming. It's on the way!

Ambulance needed - **emergency!**
Whee-oww! Whee-oww! Pull over, make way!
Lights beaming, sirens screaming,
Help is coming. It's on the way!

Tree on the track - **emergency!**
A rescue train will clear the way.
Huge crane hooking,
lifting,
shifting,
Help is coming. It's on the way!

Boat sinking fast – **emergency!**
Launch a lifeboat into storm and spray.
Ropes tossing, life preservers dropping.
Help is coming. It's on the way!

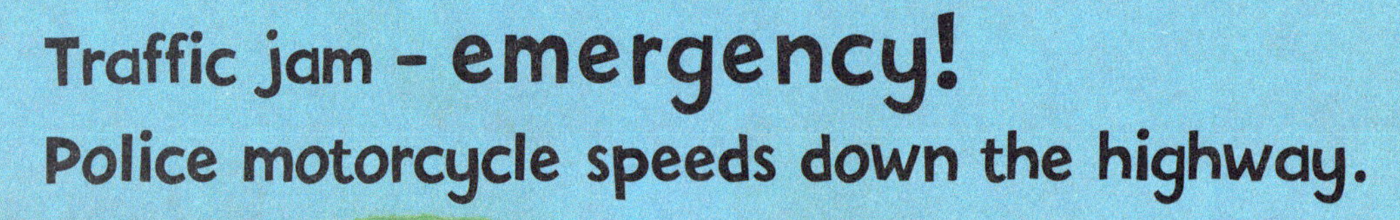

Traffic jam - **emergency!**
Police motorcycle speeds down the highway.

Zipping, revving, redirecting,
Help is coming. It's on the way!

Forest fire blazing – **emergency!**
Fire-fighting planes zoom up and away.
swoop, swoop, swooping, water scooping,
Help is coming. It's on the way!

River flooding – **emergency!**
Inflate the rafts without delay.
Supply boats tugging, chug, chug, chugging,
Help is coming. It's on the way!

Blinding blizzard - **emergency!**
Snowplow moves the snow away.
Pushing,
shoveling, tossing, tunneling,
Help is coming. It's on the way!

Fire! Fire! - **emergency!**
Fire engines race all the way.
Hoses rushing, water gushing,
Help is coming.
It's on the way!

Hiker lost on mountain – **emergency!**

Helicopter hovers and sways.

Searching, finding, side door sliding,

Help is coming. It's on the way!

Uh-oh! Accident - **emergency!**
Damaged car **blocks** the way.
Tow truck t o w i n g,
yellow light glowing,
Help is coming. It's on the way!

All quiet now – no emergency,
The rescue vehicles are all tucked away,
Ready and waiting for the next 911 call.
Then help will be coming to save the day.

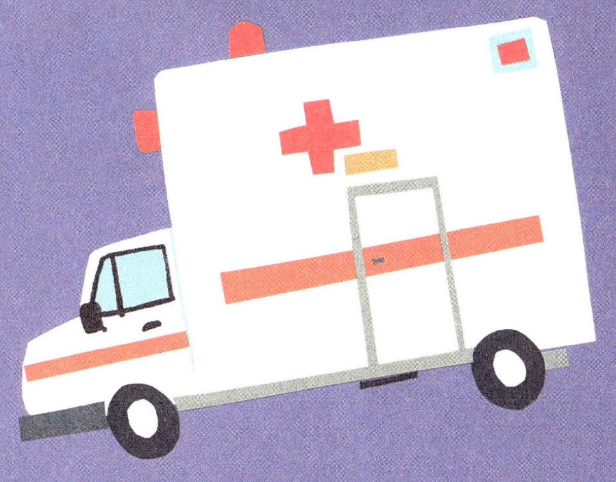